Dick Dekker

Naturalist Painter

An artist's observations of western wildlife

Western Producer Prairie Books
Saskatoon, Saskatchewan

Western Producer Prairie Books
Saskatoon, Saskatchewan

Cover & book design by Dick Dekker

Printed and bound in Canada by Modern Press
Saskatoon, Saskatchewan

Western Producer Prairie Books publications are produced and manufactured in the middle of Western Canada by a unique publishing venture owned by a group of prairie farmers who are members of Saskatchewan Wheat Pool. Our first book in 1954 was a reprint of a serial originally carried in *The Western Producer*, a weekly newspaper serving western Canadian farmers since 1923. We continue the tradition of providing enjoyable and informative reading for all Canadians.

Canadian Cataloguing in Publication Data
Dekker, Dick, 1933-
Naturalist painter

ISBN 0-88833-065-0

1. Zoology — Prairie Provinces — Pictorial works. I. Title.
QL221.P73D44 591.9'712 C80-091060-5

Contents

Introduction

Canadian artists are nature artists by tradition; they have concentrated on natural subjects, on wilderness scenery, clouds, trees, water and snow. Comparatively recently, a growing number have focused on wild animals. One has but to view the yearly shows of "Nature in Art", organized by the Canadian Nature Federation and the Royal Ontario Museum, to be impressed by the number and caliber of selected participants. Though these artists are united by the choice of subjects, their styles and approaches differ to the same degree as one person's appreciation of nature and art varies from another's. In the specialized field of bird portraiture there is a popular trend for detail in emulation of Audubon. The great master's emphasis on the multifarious splendor of a bird's plumage provided the first pictorial description of many North American species. Modern-day bird painters indulge in the ultra-refinement of detail for its own sake, seldom for the purpose of identification. At their best, such pictures convey a heightened sense of immediacy characteristic of the so-called magic realism school of painting. At their worst, these products are disciplined studies that have conditioned the buying public to equate the best in animal art with stringency of detail.

A "realistic" style of painting is well-suited to stationary objects such as flowers, mushrooms, tree trunks or a weathered barn, but it can only capture part of the reality of birds and mammals, for like man, the animal is part of nature, an entity larger than the sum of its interconnected components. The picture of an eagle clinging grotesquely to the end of a tree branch floating on the otherwise blank page of a book allows the viewer to "count the feathers", but it does not aim at the emotions. In the same way, a caged lynx lets the public ogle that which is rarely observed, but the experience speaks far less to the viewer's imagination than the briefest glimpse of the animal in its natural haunts. When seen in nature, an eagle is often just a speck in the sky. Yet the sight of one soaring over its wild domain moves the receptive viewer more than a picture replete with nostrils and scaled feet. In order to do justice to an eagle, the artist has to transmit some of the spiritual values that the bird represents. He has to convey a sense of freedom, of aloof self-sufficiency. To do so, in my view, the artist cannot detach the eagle from its natural setting. He has also to paint part of the landscape and conjure up the atmosphere of lonely plains and windy hills. The blending of the two, the animal and its proper surroundings, has to be done harmoniously and requires the complementary use of technical ability and feeling, of reason and emotion, which are the dynamic poles of the best in art.

One of the greatest animal painters of all time is Bruno Liljefors, a Swede who lived from 1860 to 1939. His monumental canvases of eagles, swans and foxes are masterpieces of fine art because of their harmony between animal and natural environment. His skill in choice of composition and color are rivalled by Carl Rungius, who painted the bears and hoofed mammals of the Canadian Rocky Mountains, but whose subjects do not cover as wide a range of animals and landscapes as Liljefors' works. The Swede captured the gloom of the winter woods as surely as the menace of a stormy sea or the bright intimate mood of the sun-filtered mini-world of butterflies and flowers. Liljefors devoted his entire life to his art. He built studios inside the northern forest and on the rocky shores of the Baltic Sea. He surrounded himself with his subjects. Tame mammals, caged birds and a multitude of skins, antlers, wings and feet provided needed reference. Such reference material is as important to the ambitious painter of animals as human models were to Rubens and Rembrandt. To enhance the lifelike proportions of their works, the great classic artists used whatever technical aids were available to them, such as calipers, elaborate grid systems or crude projection devices. The wild animal is an elusive and reluctant model. Rungius often sketched freshly killed specimen. Audubon went to work with the gun before touching pencil and brush. Contemporary animal painters routinely make use of study skins or an entirely new source of reference—photography. There is little doubt that the fidelity of the work of present-day animal artists owes much to the camera. Artists can glean a lot of useful information by looking at photographs, although outright copying is seldom possible or advisable, since the camera may distort and catch the animal in an awkward-looking pose. To the painter, a photograph is but a tool to be used with discretion in the achievement of his ultimate goal — the creation of a work of art that appeals to the emotion as well as to the intellect. In my work, I have used suitable reference material whenever available. My thanks are due to the persons who generously or unwittingly supplied some of the photographs. I am indebted also to Liljefors and Rungius, whose work provided rich sources of inspiration.

I began drawing and painting at an early age. My subjects were soon exclusively derived from the field of another hobby, natural history. The two interests have always competed for time, and still do. During spring and fall I spend all my spare time near the prairie sloughs where migratory waterfowl and shorebirds congregate. During summer, I watch foxes near their den or make hiking trips into the mountains to find their larger cousin, the wolf. There is only time for painting during the briefest days of winter when extreme cold prevents me from putting on snowshoes or skis to follow animal tracks.

One paints best what one knows and likes best. Time and again I have tried to capture on canvas my favorite scenes — a flock of geese in stormy skies, a fox on a snowdrift, or an eagle over golden foothills. I have always been fascinated by predators, avian and mammalian. Growing up in the densely populated lowlands of western Holland, birds of prey were symbols to me of all that was truly wild and natural. Since coming to Canada, I have consciously searched for those heroic animal species that were exterminated long ago in my native country. Canada is unique in the Western world in having a large share of

Canada geese are not the first migratory birds to return to the prairies, yet they are most inspirational as harbingers of spring. Their loud honking ends the long silence of winter and heralds the coming of the new season. There is a nostalgic quality to their voice that speaks of faraway places. As the flock approaches in an uneven wedge, winging slowly over the still frozen lake and the snow-flecked fields, one feels a sense of relief and wonder that the birds made it back from their journey. There is strength and comfort in the predictability of nature's seasons. The migrations of geese are an illustration of that predictability for all to see and to hear.

The first Canadas to come back to the prairies are birds that were raised locally. Northern breeders will continue to pass by for many weeks. Upon arrival on home ground, the family bond may still be intact, but soon the first-year birds will gather in a flock of their own to loaf about while the older pairs establish territories and select nesting sites. Geese are said to pair for life. The death of a partner must mean a great deal to such a loyal creature. There is at times some illegal spring shooting on the breeding grounds and I have seen lone geese cruising back and forth over the marshes, honking incessantly, as if they were searching for a lost mate. The life of the wild goose is short and punctuated by gunfire; it seldom reaches more than a fraction of the ripe old age of thirty or forty years attained by captive birds. Yet in areas where it is not molested, it may accept the close company of man, at once its greatest enemy and benefactor. Occasionally it may adopt the habits of a beggar and take bread from a hand, without ever losing the proud, dignified bearing that makes the Canada goose such an appealing personality among our native birds.

On the heels of winter (Canada geese)

A marshy corner of Beaverhill Lake is home to a small breeding population of "honkers". Although they may utilize a wide variety of artificial nesting platforms, their most natural choices are small islands and muskrat lodges. During a spring when melting snows caused the lake's level to rise nearly a meter, the geese were left with very few sites. One pair had located on a haystack in a flooded field. The competition was fierce and the gander often had to swim out to meet intruders, his neck held in a gesture of threat close along the water. His loud calls were usually sufficient to turn the trespasser about, but sometimes vicious wing-thrashing fights and aerial chases took place. In one instance, the pursuing gander forced his adversary out of the air and made it splash back into the water. The female goose took care of other enemies. When a crow landed on the haystack, she got up with gaping bill, her neck coiled like a snake ready to strike.

A second pair of Canadas had established itself on a dugout mound that was now an island on the edge of the still frozen lake, surrounded by a moat of meltwater. When I walked by, the gander always called in protest, but the female remained on the mound, lying low in the nest, her neck stretched flat on the rim of down that wafted in the breeze. One day, both geese were standing on the shore. A strong south wind had brought the melting ice to their side of the lake, and within minutes floes began covering their island, building up to a berg of arctic proportions.

If disturbed, goose nests are readily deserted. It is wise to give a brooding female a wide berth. She leaves the nest only to feed, bathe and defecate, until the eggs hatch. Twenty-four hours later, the precocial chicks are ready to follow the parents into the water, and in a few months time, they will join the adults on their hazardous journey to the South.

Arrival of the honkers (Canada geese)

The lake is ice-bound under somber, gray-blue skies that contrast with the warm browns of the dry reeds lining the shore. Over a patch of open water, nine snow geese come flying in a desultory manner; one seems ready to descend, others are undecided or turn away. They are early migrants that have out-paced the northward push of the season. Unwilling to go back and not ready to move on, these geese will probably remain in the area for some time, waiting for the weather to change.

White geese passing over the prairie provinces in spring arrive several weeks later than Canada geese, although there is usually a small vanguard of snows. By the middle of April, when the latter species begin to swell the numbers of other geese that feed on thawing stubble fields, their aggregations take on an intriguing pepper and salt composition. When there are no hunters to keep the flocks on edge, I like to watch them for signs of alarm, and whenever they rise with a great clamor of startled voices, I scan the horizon through binoculars to search for the cause of the disturbance. It usually turns out to be a small aircraft for which geese show great fear. Sometimes it is the fowl's natural enemy, the eagle. The bird may be still far off and high, a tiny dot in a wide sky, but it stands out because of its isolation. It alone moves individually on a steady course, while geese gyrate in tight flocks and straggling lines that rush overhead in an uproarious springtime display.

Despite all the commotion that a bald eagle's approach may cause, I have never seen one inflict any real damage to the geese. The only eagles that actively threaten them are golden eagles, which head for feeding flocks with powerful wing beats at impressive speeds. However, few creatures are as watchful as a goose, and the fleeing flocks always manage a safe start. One golden eagle used a different tactic. It soared up to a height where it could just be seen with the unaided eye and sailed out on flexed wings over the stubble fields on which the geese were feeding. When it fired from the sky, it fell perpendicularly like a gigantic black falcon.

Coming or going? (Snow geese)

When spring is well established on the western prairies, and every day brings the arrival of new species of birds, the mallard is too familiar to merit more than a cursory glance. However, in late winter after the skies have been empty for three or four months, the first hardy mallards to return are a feast to the eye. As the flock passes by at close range, a quick look through binoculars reveals a glimpse of the drake's glossy plumage. He is a colorful fellow: yellow bill, orange feet, green head, chestnut breast, gray belly, black and white tail, and a flash of blue on the upper wing. The drab female also has a blue speculum, which is characteristic of the species and serves an important function in pair formation; the male shows it to the duck of his choice in a ritualized gesture of amorous intent. Mallards are already paired when they come home to the breeding grounds. They make a loyal couple, jealously guarding their little bit of private space among the swimming flock, not for a moment losing sight of each other. Their activities of feeding, preening and sleeping are synchronized; when she lifts from the water, he follows like a slave.

Soon after the land is free of snow, the pair start house-hunting for a patch of vegetation suitable for hiding her fragile treasure. In about ten days, she will lay 8 to 12 eggs, their total weight approaching half her body weight. When the clutch is full, the relationship of the pair changes abruptly. While the male begins a summer of idleness, the female will spend the next 28 days hidden on the nest, leaving it only briefly for feeding and toilet purposes. She comes and goes very discreetly. If her mate is not there to protect her, she is loath to show herself to the gang of unmated drakes. Like the drakes of most duck species, mallards are violent suitors and will not hesitate to overpower a female if given the chance. So-called rape flights, in which lone female ducks are seen fleeing in terror of several pursuing males, are a common occurrence over the marshes in spring and early summer.

First mallards

By the end of winter when snowshoes or skis are no longer necessary for cross-country travel and the naturalist can choose his way from one bare patch to the next, it may still be too early to see the first migratory birds such as geese and ducks. The landscape is dormant, seemingly devoid of living things, but new color schemes are emerging and delight the painter's eye. Snow remnants create interesting patterns and intensify the warm browns, reds and yellows of soil and vegetation. Radiant willows rise like flaming torches from smoldering grass. Then suddenly, there is explosive evidence of life as a grouse bursts out of cover and whirs up and away, chuckling a guttural protest, blurred wings clattering against twigs. It is gone like a shot; there is no chance to see details, but the bird can be identified by its behavior. Instead of staying low and disappearing between the trunks of trees, as a woodland species such as the ruffed grouse would have done, it rockets upwards and jumps over the wood to drop out of view on the other side, behavior characteristic of an open country grouse such as the sharp-tailed.

A close relative of the locally extinct prairie chicken, which once occurred on the plains in untold numbers, the sharp-tailed grouse covers a wide geographic range and is periodically a common resident of open woods and brushy edges of fields that it shares with two imported fowl — the gray partridge and the pheasant. The latter two are ground feeders and may get into serious trouble if deep snow covers their food supply. Not so with the sharp-tail. It is a native species that knows how to take care of itself, feeding on the buds of trees and utilizing the insulating qualities of the snow to escape the cold. In the evening, the grouse fly into the fluffy stuff and dig a cozy hollow where they are safe from the blazing eyes of hawk and owl.

Jumping the willows (Sharp-tailed grouse)

Spring comes early to the Alberta foothills; winter snow lies still unbroken on the river flats and in corners protected from the combined strength of sun and chinook winds, but exposed slopes are usually bare by the first week of March, well in advance of other parts of the province. The sight of bare soil, even of raw mud on the cattle trails, is a relief to eyes that have grown tired of a land burdened with snow for the past four months. The scenery is splendid — to the west, distant forests rise to ice-blue mountains; to the east, undulating ranges of tan-colored hills level off to prairie plains.

Following the crest of the slopes flanking the Bow River, an eagle approaches, one wing raised slightly higher than the other, buffeted in the updrafts that suspend the heavy bird as it sails effortlessly over the uneven terrain. The vast curve of the valley behind the eagle suggests that it has come from far away. Its speed seems emphasized by the streaked patches of snow-flecked grass. The straw-yellow nape, shaped like a crown, identifies the bird as the golden eagle, the king of birds, mythical Thunderbird to the Indian, heraldic blazon of imperial power to caesars and czars.

To modern man the eagle represents conflicting qualities. To some it is still a symbol of freedom and self-sufficiency, a part of wild nature worthy of preservation; to others it is a competitor, a waster of game and a killer of livestock, a varmint that needs to be shot, trapped, poisoned or driven from the skies by means of a light aircraft or helicopter. Eagles have spread their wings over this planet since the Oligocene, 36 million years ago, and there seems no reason why such a successful species should suddenly cease to exist, were it not for the powers of a comparative newcomer, who now, at this instant in geological time, can choose whether to abuse or protect.

Coming from the mountains (Golden eagle)

Cloud shadows drift across this vast foothills scene and turn gold to brooding purple. Stark hills rise to a squalling sky that threatens with snow. At the vertex of the diagonal sweep of foreground soars a distant eagle, serene, indolent, watchful, its dominant position underlined by the sun-glowing coulee, edged by glittering remnants of winter drifts.

The necessity of achieving a balanced composition between scenery and animal required some modification of topographical features, so that this landscape, like most of the paintings reproduced in this book, does not pretend to be an authentic representation of a locality that can be pinpointed on the map. This scene is characteristic of the region west of the city of Calgary, where I have spent many glorious spring days hiking in the hills or just sitting on a lookout, scanning the sky and the land through binoculars to observe the animals, especially eagles.

When they pass over on their annual migration to northern breeding grounds, eagles can be quite numerous in the foothills. Training the glasses on a soaring eagle, others may be discovered at even greater altitudes. I have counted as many as 15 golden eagles wheeling together, corkscrewing in columns of rising air. The highest birds sailed away to be joined by others. One day, eight bald eagles were circling against a background of mountains, their regal hackles and tails dazzlingly white. On exceptional days, as many as 50 eagles may be seen to fly by, usually in pairs and singles. To the earth-bound watcher, plodding through snowdrifts on the bottomlands, laboring on slippery hillsides, or detouring around treacherous cliffs or ice, these eagles are like soaring spirits, the embodiment of boundless freedom and dreamlike movement.

Crowning the hills (Golden eagle)

The eagle has come down from its lofty position over the sunlit hills in a whistling dive, traveling so low that its shadow is almost one with the bird. Its attack is designed to take the prey by surprise. Bright-eyed, the ground squirrel has observed the danger and dashes for its nearby hole. The outcome is a toss-up, a split-second matter of life and death.

The bird pictured is a young golden eagle, probably less than three years old, as distinguished by the white, dark-tipped tail and the flashes in the wings. Young eagles appear to be less adept at catching ground squirrels than adults. They also play more or just fool around, swooping at each other in a show of aerial acrobatics, whereby the attacked rolls over on its back and presents its feet to the other bird that swoops from above. I saw young golden eagles pick up stones and drop them from a height, to recapture them in grand style. Some immatures threatened coyotes and badgers, and on one exceptional occasion, an eagle landed on a Hereford cow. The cow stopped grazing and ran down to the base of the hill, the eagle balancing on its back. When it let go and sailed away to regain a soaring position over the hill, the cow looked up at it, its usual expression of bovine wonder even more dumbfounded than usual, it seemed. The eagle involved in this rare incident showed a bulging crop as proof that its antics should not be explained by the pangs of hunger, but rather as an expression of well-being after a good meal. If the eagle had really wanted to try some beef, it could have chosen one of the many young calves which are born in the hills during the time the eagles migrate. The cow-riding episode was the only occasion when I saw an eagle take an interest in cattle at all.

A question of timing (Golden eagle)

When hunting ground squirrels in the hills, a stooping golden eagle usually approaches so fast and low that the prey has to be right in line with the bird's flight path; there is little room for braking or maneuvering. In one successful strike that I saw, the eagle seized the squirrel and kept right on going; in another the bird seemed to slam into the ground, skidding to a stop. During dull and windless days, kills are made in a less spectacular manner by low flying eagles that simply drop on their prey.

The eagle in this picture has just executed a successful hunt. It is already pulling up with mighty wing beats, a squirrel dangling in its feet. When I was choosing the elements for this painting, I was not consciously planning that they should serve an allegorical function. But unconscious forces are at play in the composition of visual art, to be probed by the viewer, elucidated upon by the critic, or pointed out by the painter. I like to think that the angles of the slope and rock slide tell the story of the eagle's fast descent. However, there are several points of view from which to approach an emotionally charged subject such as predation. Other explanations may be possible. To those who do not identify with the eagle and see no virtue or joy in its prowess as a hunter, the avalanche of stones, the result of the collapse of a rock face, may be symbolic of the squirrel's fate. Its death should be mourned like all loss of life.

Predation of one animal species on another can be a repulsive business. It is a myth that raptors or mammalian carnivores always do their job efficiently, never allowing a wounded prey to escape. However, we cannot dwell on the perceived cruelty of an eagle and forget the killing that sustains life for many other species of birds and mammals, including man. Far from gloating over the fate of their victims, or sentimentalizing their activities, one can appreciate the wild predators for what they are — creatures of beauty and superb ability that serve an important function in nature and add drama and interest for those who want to see it.

To power its wings (Golden eagle)

When there is neither carrion nor fish available, the bald eagle is forced to hunt for its food and readily turns to waterfowl, especially during late fall when the last of the ducks are massed in areas of open water. Migrating bald eagles may gather in loose flocks and I have counted as many as seventeen standing around water holes of partly frozen Cooking Lake near Edmonton. Mallards and pintails usually take off at the approach of an eagle. Diving species such as scaup and golden-eyes remain in the water and submerge when the eagle makes a low pass, thus frustrating the hunt, and the big bird planes down again to join its fellows. But sooner or later, one of them shakes its habitual lethargy and tries harder, hovering over the water against a stiff breeze, or flying in tight circles, waiting for the diving ducks to emerge. If the duck's timing is a split-second late, the eagle will snatch it up and carry it away. Some ducks that I saw fall victim were cripples, unable to react in the proper way. Others behaved unwisely; a scaup drake was the last of a group to leave a resting place on the ice, and a mallard made a mistake in judgment, returning to the water too soon. The eagle quickly doubled back and picked it up before it had another chance to flee.

When a bald eagle settles with a prey on the ice, it seldom takes long before the ubiquitous magpies turn up to wait for the left-overs, chattering excitedly. They hop around the giant and sneak up from behind to pull its tail and wing tips. The eagle seems unperturbed and may have more serious contenders to worry about — its own kind. Bald eagles have their own system of communal food distribution; if one of them catches a duck, three or four may feed on it. The successful hunter is routinely robbed by another eagle, which may have time for just a few bites until it is robbed in turn.

Wherever bald eagles are found, prey populations of waterfowl are under a steady pressure to which the infirm and the senile are first to succumb. If it were necessary to defend the role of predatory birds in nature's scheme, no better example could be found than this splendid eagle.

When feathers fly (Bald eagle and magpies)

The peregrine falcon has approached in a low surprise attack. Half a dozen black-bellied plovers notice their peril in the nick of time and rise with a cry of alarm. One of them is right in line with the falcon's flight path and appears doomed; it banks sharply in a desperate effort to get out of harm's way.

In this painting, the falcon is placed over the middle of the picture as if it moved so fast that the distance was covered before the bird was noticed. The plovers are driven into the corner. The wing posture of both plovers and falcon is somewhat unbalanced, which seems to convey a sense of action. I have witnessed similar scenes in nature many times, but the vehemence of the moment never fails to grip and excite. The eye becomes insatiable for falcons; I have sought them at every opportunity.

The peregrine is the fastest creature in the world, whose level flying speed is reported to be around 100 kilometers an hour. It can quickly double or triple that velocity by utilizing the force of gravity during the so-called stoop — when it pulls in its wings or holds them flexed close to the streamlined body. The falcon falls like a meteor, so fast that the human eye perceives it as a streak. An accurate measurement of the maximum speed obtained is hard to come by, but a convincing testimony was published in a classical work on raptors, titled *Life Histories of North-American Birds of Prey,* by A. C. Bent. He reported that in 1930, an expert aviator was cruising in his small pursuit plane when he saw a flock of ducks flying far below him. Thinking of gaining some experience in diving at a moving target, he opened the throttle, increasing his speed to 280 kilometers an hour. Glancing at the wing tip to see how much vibration his dive was causing, the pilot spotted a peregrine that was stooping at the ducks. It shot by at a speed far in excess of that of the plane and hit one of the ducks that fell from the sky, apparently lifeless.

Panic among the plover (Peregrine falcon)

The peregrine is reputed to be a near-infallible hunter, capable of catching its prey at will. The handbooks state that this falcon usually captures a large bird such as a duck in midair, dealing it a blow from the talons that sends the victim to earth, stunned or killed outright. Such spectacular hunting methods may be rarer than is generally believed. I have watched 700 hunting flights by peregrines of which 60 were successful; in all observed instances the prey was seized in a straightforward manner and simply borne down to earth.

Far from being infallible, some peregrines, especially the young, may chase one bird after the other and fail to make contact, frustrated every time by the quick-witted responses of the prey. A peregrine may be faster than the birds it preys upon, but the latter are usually more maneuverable and seem to know exactly what to do when attacked. An often-used escape tactic is to drop like a stone and take cover on the ground or in water. The timing of this routine is so astute that it may seem as if the prey was actually hit and fell to earth lifeless, thus giving rise to the false notion that the peregrine kills at will, often striking two or three victims in a row and letting them drop without an attempt at retrieval. The conscientious observer can be sure of a hit only when feathers are seen to fly or when the victim is picked up and found to be wounded or dead.

The peregrine in this painting is a young bird as distinguished by the brownish dorsal coloration. During the second year the brown feathers are molted out and replaced by blue-gray ones. This young falcon, its long sickle wings arching high over its back, is chasing a party of pintails, three drakes and one duck, that have reached the lake in time and will splash down in an instant just as the falcon closes in. A spray of water will shoot upward and may hit the hunter, cooling its fervor.*

* Detailed accounts of my observations on peregrines were published in the *Canadian Field Naturalist*, number 3, 1979 and number 4, 1980.

Futile chase (Peregrine and pintail ducks)

Soon after the warm colors of grass and soil replace the deathly white of winter, and pools of meltwater mirror skies that vibrate with migrating ducks and shorebirds, I search for that brilliant product of avian evolution, the peregrine falcon. The moment of discovery is like the finding of a lost jewel. Through binoculars, there is no mistaking its silhouette — wide shoulders tapering to long wings and tail, bullet-shaped head contrasting darkly against a whitish chest. The falcon stands very erect and is continually scanning the country with quick turns of its head, briefly focussing on me as I approach slowly and find a seat against a stone or post some distance away. A 20-power scope allows me to study its striking features. The bold moustachial lobes form an impassive mask that frame the great luminous eyes, their expression docile, magnanimous, belying the menace that emanates from the clawed feet clasping the post.

Peregrines sit on posts in open country for hours at a time, baked by hot sun, buffeted by winds, pelted by hail or drenched in day-long rain. During a torrential thunder shower, I watched a peregrine flap its wings for minutes on end as if it were flying through a waterfall, but it stayed on its post, ruffling its feathers and preening contentedly after the storm had passed.

In the hope of seeing a peregrine begin to hunt, I have spent hours at a time watching them as they sat perched. The boredom of the long wait was immediately forgotten as soon as the bird raised its wings in the deliberate, almost sensual manner characteristic of the species. For a second or two prior to launching, the long wings are held fully spread at an angle slightly above the horizontal. Then it drops off the post, falling to just over the ground before climbing into the wind on swiftly sculling wings.

A life of leisure (Peregrine falcon)

Because of their slender wings and snappy flight, falcons are generally not considered as soaring hawks. However the peregrines that I watched migrating through Alberta on their way to and from northern breeding grounds soared if conditions were favorable. On sunny and windy days they spent hours aloft on spread wings and tail, rising so high overhead that they eluded the unaided eye and were just visible in ten-power binoculars. Some entered the base of cumulus clouds that top thermal currents; obscured by fog, they suddenly faded from sight at altitudes that might exceed 2000 meters. During their corkscrewing ascent, the falcons drifted downwind, but once they stopped circling, they could choose any direction, sailing away at great speed on flexed wings.

Before leaving on migration, some falcons needed a bite to eat. Spotting a target far below, they pulled in their wings and fell like a thunderbolt in a breath-taking stoop, revolving so that the sun lit up their underside like a flash of fire. Swinging level just before reaching the target and overtaking it at tremendous speed, they tried to seize the prey — a low-flying duck or a shorebird that flushed from the ground.

This was their favorite way of attacking flocks of feeding or resting sandpipers. After a foiled hunt, the falcon swooped up and might make an additional pass at the fleeing bird, perhaps pursuing it for some distance, but more often it flew upwind and began soaring again, regaining its lofty position until ready for the next stoop. If it were successful in capturing a sandpiper, the falcon landed to eat it. The smaller peregrines were often bothered by red-tailed or Swainson's hawks that tried to steal the prey, forcing the falcon to transport its booty to safer locations. Once I saw an adult male outclimb several pursuing hawks and begin to soar, reaching great altitude where he drifted and consumed his prey. His tiny silhouette, a speck in the blue, emitted plumes of feathers like puffs of white smoke.

Soaring serenely (Peregrine falcon)

Sandstone cliffs along the Red Deer River are the home of the prairie falcon. The female lays her eggs on a sheltered ledge without using any soft material as underlay. When disturbed by the approach of humans, the pair protests emphatically, screaming at the top of their grating, staccato voices. It is best to stay at a respectable distance and view the events through binoculars.

While the female returns to the nest, the male stands guard on one of his favorite pinnacles. The rock face below him is streaked with white excrement as proof of his regular occupancy. He is a proud figure of a falcon, the picture of vigilance as he scans the scenic valley with quick turns of his round head. Whenever another falcon comes too close and violates the airspace around the cliff, he sallies forth on a swift mission of defense. He also keeps magpies and crows away and thus involuntarily protects the colony of cliff swallows that built numerous mud-made nest cones against lower walls of the cliff. At another site, not defended by falcons, I saw a crow take several young swallows, despite the protests of hundreds of milling parents. At a falcon cliff, other enemies of the swallows, such as sharp-shinned hawks and merlins, are kept away too. The falcons themselves never seem to bother their little neighbors. They hunt different prey; a few open-country birds like larks and sparrows, but mainly ground squirrels that are carried to the eyrie as food for the female and chicks.

By the time the downy young are half-grown and beginning to grow feathers, both adults spend endless time soaring and drifting high over the fields and pastures, launching their hunting flights in beautiful stoops that carry them far out of sight.

Over the eyrie cliffs (Prairie falcons)

The slant of the shoreline seems to emphasize the merlin's speed as it dashes by, sharply silhouetted against the water, pointed wings swept back, clawed feet lowering to seize one of three sandpipers that have just flushed from the mud. The animated postures of both the little falcon and the sandpipers convey the urgency of the moment; life or death is a matter of centimeters . . . .

The merlin is a perfect subject for the student of birds of prey; it is spectacularly swift and agile, inhabits open country and is neither very rare nor particularly shy so that it can be readily observed. I have often seen it hunting and recorded 285 occasions on which it made one or more attempts at seizing a small bird. In 29 attacks the prey was killed, constituting a success rate of nearly 10 percent. Thus, despite its great speed and dexterity, the merlin catches only one out of every ten birds it chases. But my figures do not necessarily prove that the observed merlins were unable to catch the other nine birds; perhaps the little falcons were unwilling or not sufficiently hungry to try harder. Motivation is an important element deciding the outcome of hunting attempts. Adult male merlins that I saw foraging for their mate or brood on the nest achieved a success rate of 35 percent. No doubt, adult merlins are better hunters than immatures.

During their first weeks of independence, young merlins indulge in chasing sandpipers, but are usually outflown and out-maneuvered. To hunt these fast little birds with the greatest chance of success, the merlin learns to capitalize on surprise, coming in low and attempting to get as close as possible before being noticed. If the strategy fails or if the prospective prey dodges, the hunt may be decided there and then, with the merlin alighting on a post as if to contemplate its next move when it may not give up so easily. Another time the merlin will keep up the chase, swooping repeatedly, until the prey is caught or escapes into a cover of reeds or bushes. Several hotly pursued sandpipers that I saw drop into water to escape were picked up from the surface by this versatile little predator.

Action and reaction (Merlin and least sandpipers)

The marsh hawk or harrier is easily the most visible bird of prey as it courses low over open country in an active search for food. Particularly during late summer when the new generation is on the wing, numerous hawks can be observed locally. Their restless movements near and far create a constant vibration over the flat horizons of Beaverhill Lake, the locale of this painting.

Young harriers are always on the go, hunting or playing, fighting and mobbing. If their course follows the lake shore, they are accompanied by the rising of ducks that rush to the water and dive if pursued by the hawk, long legs dangling. The fowl are seldom in real danger, but I saw a sick teal seized by the head and dragged onto land to be fed upon by six or seven harriers in turn. The marsh hawk's principal prey is the meadow vole. Where this small rodent is abundant, the harrier thrives and may be seen displaying characteristic U-shaped dives and ascensions. During times of food scarcity such nuptial sport is rare and successful nesting is jeopardized. The adult birds are then forced to change their diet. The large brown females may take a more serious try at waterfowl. I saw a hawk, up to her belly in water, as she tackled a coot that submerged or was pushed under by the weight of the attacker. She let go when the coot's mate rushed up in a threatening posture.

The adult males, recognizable by their light gray plumage and black wing tips, regularly hunt sparrows and small shorebirds in a style very unlike a harrier — winging swiftly just over the grass, they attempt to take the prey by surprise as it flushes just ahead. Their success is not guaranteed; of several dozen tries only one was a hit. In a lightning quick somersault, the adult male captured a small bird that rose from the grass. He kept right on going for he was immediately pursued by an immature harrier. I watched through binoculars as they flew far out and high over the lake. Two kilometers past the start of the chase, the uppermost bird dropped a tiny morsel which was caught in the air by the lower one. It turned and sailed back to land.

Winging ways (Marsh hawk)

Sandpipers live on the edge of water and land. They flash white undersides when they fly. Their brown backs fade into the sand upon landing. They are ethereal creatures, evanescent like the gleam of sky reflected in a wavelet that washes the shore. Coming and going by night, they make globe-spanning journeys along the Pacific and Atlantic coasts or via a direct route through the heart of the continent. They are birds of ocean shores, of tidal flats and surf-washed beaches, where the perpetual cycle of ebb and flow guarantees a daily replenished supply of food. Most of them breed in the arctic along shallow tundra pools, rich in small crustaceans and insect larvae. During spring and fall, they may stop over at suitable localities far inland.

Among birdwatchers, Beaverhill Lake is renowned for the masses of migrating shorebirds that can be seen along its shores. This shallow prairie lake is so large that strong winds create a tidal action by pushing water away from the upwind side. Local weather conditions, such as periods of drought, can also influence the availability of mudflat habitat that sandpipers need. However, here such habitat is temporary and subject to constant change. The lake bottom exposed by dropping water levels soon dries out, grows up in grass, weeds and willows, and eventually becomes unsuitable for sandpipers.

During times of high water, shoreline vegetation is inundated and killed as it slowly rots away. When the lake recedes again, the cycle is repeated. Severe drops in water level are also essential to retard the spread of aquatic vegetation that would otherwise clog such shallow lakes or border them with dense stands of reeds and rushes, rendering the habitat unsuitable for these delightful little denizens of the water's edge.

It is a challenge to the naturalist to try and understand the subtle yet profound interrelationship between weather and habitat, vegetation and bird life. The beauty created by the interplay between clouds and puddles, water and mud, is an inspiration to the painter.

Between clouds and puddles (Sandpipers)

The greater yellowlegs is a tall, slender shorebird that nests near bogs and muskegs in the northern boreal forest and the mountains. While the female broods the eggs in a mossy depression, the male stands guard on the very tip of a spruce to warn of dangers, be they animal or human. Upon hearing its staccato calls ringing among the trees, I wonder what kind of animal caused the alarm, whether it is just a moose running away, or a bear meeting me on the wilderness trail ahead. The call of the yellowlegs has a nervous twang that bespeaks emergency and panic. The bird always seems to be alert and on edge, reacting with explosive intensity. Its life quickens to a pulse we never reach nor comprehend. Like all birds, it is old and wise to its world in a matter of months after struggling out of the egg.

Once the worries of raising a brood are over, greater yellowlegs gather along shallow muddy water on the prairies, joining their smaller cousin, the lesser yellowlegs. Tranquil late summer afternoons at Beaverhill Lake are pierced by the querulous calls of these long-legged sentinels, their bulging eyes ceaselessly scanning land and water for foes, real or imagined. If I sit down, the pastoral quiet is soon restored and the birds may approach closely, allowing an opportunity for study. The greater yellowlegs is a hunter that stalks its prey like a man holding a spear. Walking gravely through the shallows, lifting its long yellow legs high to avoid drag, the bird stirs up aquatic insects and freshwater shrimps, seizing them in a quick peck. It may sprint after one and stab repeatedly. At times, it captures a minnow or stickleback and carries it to land to shake and smash it until it can be swallowed, or it may catch a large water beetle and hammer it to death, pulling out its legs one by one, each taking minutes of wrenching and tugging. Killing, and in our eyes, cruel treatment of prey, is not confined to the predators. If we condone such activity by a shorebird, we cannot condemn the falcon that suddenly shatters the pastoral calm and falls like a bolt of lightning out of the blue to pursue the yellowlegs that flees for its life.

August calm (Greater yellowlegs)

In ancient times, when every living creature and even inanimate objects were believed to be possessed by spirits, the owl was universally feared as a symbol of portent, a harbinger of death and destruction. Its magical powers were enlisted by shamans and sorcerers to ward off the forces of evil in an age-old tactic of fighting fire with fire. Of course, little is left of those superstitious attitudes in the folklore of our part of the world. Yet to many people, this creature of the night that is so admirably adapted to its role in nature, still holds a special fascination and continues to be widely used as a subject for artists of all stripes.

Quite unique among birds, the owl's round face seems endowed with human expression. To suddenly discover one sitting quietly in a tree, wide eyes fixed on yours in a solemn unblinking stare, is to be jolted with the uneasy feeling of having been spied upon. It is this startling moment of meeting the owl's enigmated gaze that I have tried to illustrate in this picture. Instead of concentrating on the details of its delicately patterned plumage, I have represented the owl as belonging wholly to its treed environment, cryptic in the tangled branches and the mottled bark of the spruces.

Belying its indolent and docile daytime mien, this great-horned owl is a hunter and killer par excellence, although its deeds are usually obscured by darkness and seldom witnessed by man. During winter, it may carry its nighttime kill along, holding on to a half-eaten rabbit or other large prey until its appetite returns. The great-horned owl shares this "tight-fisted" habit with the snowy owl, which I saw transporting the remains of hares and mallards, refusing to leave them behind even when not hungry, thus preventing their loss to scavengers. Such thrift is quite unusual among raptorial birds and appears to be an adaptation that may help these northern owls to cope with lean times.

Watching the watcher (Great-horned owl)

The first snowy owl I ever saw was sitting on a stone in a frozen marsh, watching me intently as I approached slowly for a closer look at this stately visitor from arctic tundras. When it flew, it seemed as large as a swan without a neck, but its flight was more buoyant, the solid body rocking lightly from the thrust of powerful wings.

Since that first encounter, familiarity has dulled the excitement of further sightings of this beautiful owl that is a common migrant in central Alberta, especially during late fall and spring. From one point of land I once counted 26 of them sitting on posts and haystacks, surveying the snow-covered fields that abounded in meadow voles. If well-fed, they may rest for hours. Observing them is usually a cold and boring business and one day I gave in to the temptation to catch one of these aloof creatures. After dusk, one came to a trap baited with a pigeon, and I took it home for a closer look. Inside the house, the owl flew to a bookcase and regained its serene composure, looking like an artifact of stunning detail and texture. Its eyes were pools of liquid gold, shaded by lids of hoar-frosted lace. When it closed its eyes to slits, the bird became a bundle of feathers that covered all but the tip of the heavy bill and the needle-sharp nails. It was certainly well equipped to survive extreme cold.

I soon released my captive and to all appearances, it was none the worse for its experience, but its former tolerance of passing cars and snowshoeing bird-watchers was gone. The cautious regard for humans which the bird acquired during its captivity may some day prove life-saving, for the lack of shyness of these northern owls makes them easy targets for the vandal's bullet. Eskimos have culinary reasons for wanting to shoot the "ookpik". They may try to lure it within rifle range by pulling a piece of fur on a string behind them. A variant of this method was used by a Calgary naturalist to investigate the visual acuity of these owls that hunt by day as well as night. He cast his bait from a car window by means of a fishing rod and found that the owls came to attack the mink-sized "prey" from distances of up to 1600 meters.

Owlish reflection (Snowy owl)

The low light of the rising sun brings out the multi-shaded greens of forested hills and the jagged relief of mountains beyond. The dry grasses of the meadow gleam yellow, a hint of frosty gray in the hollows. Three wolves have just emerged from the spruces and halt in the open, presenting a thrilling sight that I observed one September morning in Jasper National Park.

They were large wolves belonging to a western race that averages thirty to forty pounds heavier than eastern Canadian races. One was evenly black; the others were dark only on their thickly furred back, and a creamy white on their belly and legs. The color pattern of the head presented a striking mongolian mask in which the eyes shone amber, their gaze inscrutable without a hint of menace. The black one's eyes looked like pale gold in its grizzled face, the expression inquisitive and harmless as a child's. But their docile looks belied their reputation; these wolves were the masters of this valley, the ultimate predators that exacted their pound of flesh from five species of ungulates: moose, elk, deer, sheep and goat.

In 1965 when I first visited the valley, wolves were very scarce in Jasper Park. In preceding years they had been poisoned on adjacent provincial lands, while hoofed mammals were numerous and had outgrown local food supplies. The wintering ranges of the park were badly overgrazed; the damaged bark of poplar trees was silent evidence of the fact that elk were on a starvation diet. To reduce the herds, wardens shot several hundred cows yearly. Biologists feared that mountain sheep populations would soon crash as a result of disease induced by malnutrition. However, before their prediction could come true, the predator policy of the Alberta government changed and the poisoning all but stopped. Wolves made a big comeback and began to cull the elk herds, preying heavily on calves. Grass on the meadows recovered slowly and dense thickets of poplars are now appearing where none had a chance to grow before.

Wolves of wilderness valley

A good method to use in studying the habits of wolves is to follow their trails in snow. When traveling single file, the animals habitually step in each other's footprints so that a large pack may leave only a single spoor. One trail that I followed split ten times, proving that the pack consisted of at least eleven members.

In places where hunting action has taken place, the unraveling of signs is complicated, especially if there are several prey animals on the run. A pack of wolves chasing a couple of deer can leave behind a confusion of tracks. The charging wolves create long oval troughs indicative of extensive displacement of snow. Deer, on the other hand, experience minimal drag by virtue of their high, bounding gait. In deep soft snow, the wolves stay close to big spruce trees wherever possible, so as to benefit from the lesser snow depth at the base. Attempting to intercept the deer, they may crash through bushes and tangles of deadfall, but more likely than not, the chase will leave them panting and with nothing to eat. So the trail leads on . . . mile upon mile.

The wolves' chances of success are better if they travel the ridge tops and attack on a downward course, plunging through deep drifts in which their prey may flounder. Steepness of terrain may inhibit the deer's jumping and slow it down so that the wolves have time to close in. At the site of a deer kill, usually very little edible material is left. The stomach contents, a frozen clump of fiber, lie on a mat of hair among a few splintered bones. By the time I arrived at the site, the wolves had already left, probably on their way to distant hunting grounds.

Following the leader (Wolves)

Ravens and wolves are often seen together for an obvious reason; the relationship between them is an example of commensalism — one species feeds on the leftovers of the other. The raven is clearly the beneficiary, although it may just be possible that on occasion the clever bird assists the wolf in locating prey animals. A biologist who followed wolves by aircraft saw ravens fly along the trail as if they had the same purpose in mind as he — to catch up with the pack. Wolves tend to travel by circuits, and habitually follow topographical features such as lakes and ridges. Perhaps ravens have learned to locate certain packs by flying along their often used runways.

When following wolf trails and coming across the remains of a kill, I was sometimes puzzled by finding fresh wolf prints on top of the aged tracks of the pack. I learned that the fresh tracks were made by so called "cold meat wolves", social outcasts that shadow the pack and scavenge on leftovers. The lone wolves are unable to kill large prey such as moose and may depend on the pack for a living, although they have to keep their distance for fear of harsh treatment.

The big dark wolf in the painting is a loner that I saw crossing a melting lake in Jasper National Park. The wolf's big feet kicked up slush, while the ravens hopped along in short flights. At a distance, when size is an unreliable criterion for identifying certain animals, a wolf may look very much like a coyote, especially if its color is gray and brown. In such cases, it may be useful to keep in mind that a wolf has relatively large feet, whereas a trotting coyote may convey a dainty impression as if it were moving on tiptoe, like a ballerina.

Comrades (Wolf and ravens)

Towards evening, the quiet places all over the west come alive with the sound of "il soprano". His boisterous yapping and high-pitched choruses have the excited ring of children's voices at school's end. The calls carry across the land in a wave of canid music when neighboring coyotes respond, communicating their whereabouts and perhaps other information only they understand.

On much of the southern prairies and foothills, where kit foxes and wolves have been exterminated long ago and where the red fox has a sparse distribution, the coyote is the only wild canid. This adaptable species has survived despite much persecution. In the past, that persecution was based largely on the notion that the coyote was a threat to game animals and livestock, or that wild canids were the only carriers of rabies. To prevent the spread of this feared disease, the Alberta government organized a large-scale extermination program during the early 1950s when an estimated 117,000 coyotes were killed, mainly by the use of poison. In recent years, the dominating factor in the pursuit of coyotes has been greed. Fur dealers, backed by European demand, have put a record high price on the animal's head. Both coyote and fox are hunted from fall to spring by all legal and illegal means. I found it particularly distressing to discover snares and traps in the entrance of dens as late as the end of March when the female coyote and the vixen are carrying the unborn. These trappers harm their own interest by destroying the source of the golden fur.

No doubt some coyotes will survive and expand their population soon after the pressure eases, perhaps by more effective enforcement of protective legislation, or by random factors such as plummeting fur prices or changes in the cost and availability of gasoline which would curtail modern man's unprecedented mobility. Then pairs of coyotes will once again be a common sight as they trot over the open hills, stopping when their neighbors call to join them in the daily sundown serenade.

Hill-country callers (Coyotes)

A white-tailed buck, the picture of pride and poise, steps out onto the snow-covered creek bank, while a doe trots down the slope, her flag raised as a signal of danger. Perhaps she is fleeing from foes lurking in the woods. Deer, especially the young, have many enemies, but adults can defend themselves surprisingly well. One day, I saw a doe confront a coyote that tried repeatedly to enter a stand of spruce on a point of land jutting out into a river. Each time, the deer charged and the coyote fled in a hurry. No doubt the doe was protecting a calf hidden among the trees. Groups of adult deer usually ignore coyotes, even when they walk right by. However, under some circumstances, perhaps dependent upon the availability of other food or on the condition of the deer, coyotes are a serious predator. I saw a similar lack of concern in the face of danger displayed by mule deer, a Rocky Mountain species characterized by very large ears. Four or five of these deer were daily visitors on a meadow in Jasper National Park, which had been home to a family group of eleven wolves for several weeks. Apparently the deer did not find it necessary to vacate their meadow and fled only when the wolves were close. This nonchalant attitude towards predators may be a good indicator of the deer's confidence in its chances of escaping. Its speed and the size of its jumps are indeed astonishing. How their spindly legs stand up under their body weight and absorb the impact of those jumps seems a miracle; a deer's leg is another example of nature's functional design that combines strength with lightness of material.

To see a group of white-tails bound away, leaping and prancing, their large flags waving, is to smile with cheer and admiration. This deer is without question a universal favorite among our native wildlife. Because of its tolerance of danger and its ability to cope with enemies, it can live close to man, its admirer and most treacherous foe. Among experienced hunters, who know the elusive white-tail better than anyone else, the buck has a reputation for wilyness and adaptability that makes him the subject of many a fondly told tale.

Pretty pair (White-tailed deer)

During the days when the last of the lobos roamed the western cattle ranges, they reportedly snapped off the tail of many a heifer or steer. This tidbit of information, recorded in S. P. Young's classical book on North American wolves, points to the significance of the absence of long tails on wild ungulates such as moose, wapiti, sheep, deer and antelope. If these animals had more than just a stubby appendix on their rear end, they might risk the chance of being "held back" by pursuing predators, which themselves have long tails that function as rudders and facilitate fast maneuvering. The white-tailed deer seems an exception; this prey species sports a surprisingly large tail, but if danger threatens, it is carried vertically over its back, well out of the way of pursuing wolves. Apart from its function as a flare, the raised flag of the white-tail serves the same purpose as the light areas on the rump of other hoofed mammals — it is a guiding beacon for following fawns.

The white-tailed deer is a dynamic animal that has benefited from man's protection; in many areas of the continent there are now more deer than before the coming of the settlers. Originally an eastern and southern species, the white-tail has expanded its range to the north and west. In recent years, I have met a few of them in the mountains of Jasper National Park where they mingle with their western cousin, the mule deer. A few cases of interbreeding have been recorded. Apparently the tail of the offspring is neither as large as the flag of the white-tail, nor are the ears as big as those of the mule deer. As these hybrids are fertile, one wonders whether the deer of the future will display characteristics of the two.

On the painting, a doe stands transfixed on the edge of a clearing. Snow-laden spruces reflect the warm glow of the evening sky.

Winter white-tail (Deer)

One winter's day I was skiing on a frozen creek in the Rocky Mountains, when a cow elk suddenly rose from her bed on an island and jumped down the steep shore onto a patch of ice to slip and fall. She tried desperately to get up, but one of her front legs appeared to be broken and with the other she could not find solid footing. She managed to push herself against the bank, where she stayed until I returned an hour later. The snow around her was packed by fresh coyote tracks.

The breaking or crippling of a leg does not necessarily prevent an elk or deer from continuing its daily routine. Once I saw a white-tail run uphill at great speed despite a lame foreleg that dangled and swayed by its side. Naturally, such a handicap would render a deer more vulnerable to its predators; a wolf would make short shrift with a three-legged deer or a wounded elk which is unable to get up from the ice. At the time when I caused the accidental fall of the elk cow, wolves had been all but exterminated from the Alberta mountain forests and the unfortunate elk was left to the coyotes, which do not have the power to dispatch such a large animal quickly. If the cow did not stand up soon, she would be eaten alive. . . .

Unable to assist the elk or put her to death, I drove to the nearest forestry station, arriving as darkness fell, but to this day I am not sure whether the ranger went out to find the elk. Every time I recall the episode, I relive the agony of having to leave the pathetic creature to a fate that it shares with countless other elk and deer: the fate to die a violent death. Rather than dwelling on this terrible truth, I prefer to recall other experiences with elk, such as meeting a herd on a colorful mountain slope one September day. One by one, half a dozen cows broke out of the trees and stood watching, while a bull bugled behind them. Presently, he appeared and thrashed the ground, throwing clods of earth over his back before trotting off after the cows, his heavy antlers festooned with moss.

Watchful wapiti (Elk cows)

The awesome scale of the Rocky Mountains tends to overwhelm visitors from the plains who may drive mile after mile without giving the peaks more than a casual glance. However, as soon as their eyes focus on an animal, such as a moose standing in a roadside lake, they may slam the brakes and endanger traffic safety just to have a second look. The small figure of the moose creates a focal point that brings the great scenic spectacle down to the size of living things, thereby enhancing the perspective and grandeur of the mountains in the background.

While other hoofed animals of the Rockies, such as wapiti, sheep and goats, have gregarious habits, the moose is nearly always alone. It is the least sociable of the ungulates, a self-contained solitaire with exclusive features, perfectly adapted to its myopic existence in the northern woods across the top of the world, from Scandinavia through Siberia and Alaska to Labrador. Its long legs allow it to wade through bogs and into lakes and rivers, where it feeds on aquatic vegetation, a favorite summer dish. It may dive to a depth of several meters to uproot water lilies, leaving a raft of bulbous tubers floating along the shore. While standing in the water, it may also find relief from the torment of flies and mosquitoes, but on calm days the pests cannot be shaken off so easily.

During a canoe trip along the wilderness highway of the Yukon, I encountered a submerged moose of which only the head was showing under a balloon of insects. The dense swarm followed the moose as it swam across the river. The animal rose up the shore like a prehistoric monster from the deep, water streaming from its black body, the dewlap dangling under the curiously long head with the pendulous nose. Moose may also enter water to escape predators. The cows usually seek out secluded spots along a lake or on an island to serve as a nursery for the newborn calves.

Moose mountains

The settlers who first set eyes on the ridiculously spindly long-eared bunny of the southwestern deserts called it an unflattering name — jackass rabbit, which has since been shortened to jack. Although this species is not a rabbit at all but a true hare, the scientists have stuck to that inaccurate label and have named the northern race, which is less grotesquely slender than its southern cousin, the white-tailed jack rabbit.

Our jack strikes a grand picture, particularly in its white winter coat as it bounds stiff-legged across the snowbound fields and sends the dry powder smoking and glittering at every leap. These large hares may weigh up to twelve pounds, as heavy as an average fox. They are hardy critters that only occasionally go underground in a badger burrow, relying on their cryptic coloration and great speed to escape from enemies such as coyotes, eagles and owls, as well as man on his snowmobile. Zig-zagging at high speed, the jack can throw off all but the most determined pursuers. One bitterly cold day, I saw a gyrfalcon repeatedly swoop at a hare that met every attack by jumping upwards, meeting the hawk in mid-air, aborting its plunge. Eventually, the jack took shelter under a barbed wire fence from which it refused to budge.

No doubt many a hare loses its life during the course of winter, but a surprisingly high number may survive to become much more visible in spring when the bunnies go courting. I have seen as many as a dozen of them together, all taking part in a rough and tumble mating ritual of intense preoccupation. Some of them passed by closely, unaware of my presence, their noses to the ground as if they were following a scent. If they flushed another hare, probably a female, from its hiding place, they chased it through mud, slush and water, while other hares joined along the way.

White on white (Jack rabbit)

The lynx is a secretive denizen of the northwoods where it leads a solitary life of feast or famine. During years of plenty when its main prey, the snowshoe rabbit or varying hare, is at the peak of its ten-year cycle of abundance, the cat can be locally common, yet seldom seen and hard to get to know. In snow, its large fuzzy feet leave a characteristic spoor of shallow impressions, round and closely spaced. But as soon as the animal enters a brushy rabbit patch and chooses to follow the well-packed runways, its tracks are soon lost, especially if it goes bounding through the brush on an erratic course in pursuit of a rabbit. Once I found the freshly-gnawed hind quarters of a hare lying on a circular area of trampled snow, but the lynx was nowhere to be seen and its tracks were too indistinct to unravel the hunt that had taken place.

During summer, the lynx does not leave any sign of its presence and one is lucky to catch a glimpse as it crosses a road or forest trail. In the foothills, I once saw it emerge from the woods to snap up a ground squirrel in a quick dash and carry it back into a tangled forest.

My most memorable meeting occurred on a winter day when the alarm barks of coyotes caused me to stop and glass the edge of the woods in which I had just seen three of the wild dogs disappear. Presently, a lynx emerged and strode into the open pasture flanking a creek. I sat down and waited for it to come closer and eventually it passed at a distance of about 20 meters. When I sounded a mousy squeak, it turned and walked even nearer until it halted and fixed me in a sphynx-like gaze for a few seconds. It turned around without haste or hint of emotion, dismissing me as if I were a stump or stone.

Apart from its aloof, enigmatic expression, the single-most outstanding feature of this strange and fascinating creature was its feet: shapeless clumps of fur that looked like dust mops. It was hard to imagine that those sloppy slippers contained a set of deadly, retractable weapons.

Silent paws (Lynx)

One of the rarest mammals on the prairies and in the Rocky Mountains of Alberta is the otter. It was for this reason that I was thrilled to find signs of its occurrence in Jasper Park.

During a spring of exceptionally low water levels, the stream connecting a lake with the Athabasca River had all but dried up, so that traveling otters were forced to run instead of swim. Their characteristically askew track patterns were preserved in mud alongside the stream. The prints were of various ages and sizes, indicating that this route had been used regularly. During subsequent checks of the area, I located other signs on the banks of the lake. On some spots, the grass was flattened, as if the animals had rolled there often. Nearby were accumulations of flat, scaly excrement that gave off a peculiar but not unpleasant smell, a fishy kind of perfume. Holes and burrows that looked like potential dens riddled some steep banks. In front of one burrow, the mud was pitted with tracks. When I stuck my head into the wide opening, a throaty growl coming from inside made me withdraw in a hurry. Just after I had left, a pair of Canada geese gave an explosive alarm that seemed to announce the otter's emergence from the den, but I never saw it at all.

Otters can be very secretive and may swim underwater for long distances to emerge with only the top of the head showing. One fall day, I surprised an adult with two cubs, all curled up together on a muskrat lodge. The parent dived with a splash and was not seen again, but the young cruised around the canoe for awhile, snorting and sneezing, like miniature Nessies. Their long necks were held vertically, well above the surface. The round, impudent faces bristled with whiskers.

These Jasper Park otters have all but vanished. The lake is used by increasing numbers of tourists. On the highway that parallels the shore, at least two otters were killed by traffic. A safer habitat for them is provided by Prince Albert National Park, where I found inspiration to do the accompanying painting.

Spring thaw (Otter)

Early winter, before snow covers the ice, is a good time to explore the cozy world of the muskrat, a common inhabitant of shallow lakes and sloughs, admirably capable of coping during the long freeze-up to come. Like the beaver, the muskrat builds a lodge, hollowing out a mound of cattails or bulrushes to create a living space just above water level, protected from the outside world by a rock-hard roof of frozen plants. Unlike beavers, muskrats do not lay up food stores but travel to winter feeding grounds via well-established routes that can be traced by air bubbles. It is believed that these pockets of exhaled air are replenished with oxygen from the water and that the rats use the recycled air whenever needed. At the end of the passage ways radiating out from the lodge are the so-called push-ups, little domes of plant material and mud that cover a round hole and prevent it from freezing. Additional insulation is provided by snow. If ice forms in the holes, the rats will attempt to break it up, maintaining their ingenious feeding and breathing stations by frequent visits.

These amphibious rodents are in fact so successful in exploiting their marshy habitat, summer and winter, that they periodically overpopulate their living space, despite an army of enemies. During such times of stress, intraspecific fighting, emigration and disease may combine to reduce the population to a minimum.

The plump little muskrat is a vegetarian known to eat a few frogs and clams, but one day I observed one that had developed a taste for fish. Apparently it was exploiting a very dense concentration of minnows and sticklebacks. I saw it dive 21 times within an hour and every time it emerged with a small fish held sideways in its mouth. After it had clambered onto land, it transferred the catch to its front feet, holding the tail upwards and consuming it like a monkey would eat a banana. I could tell by its movements that the fish was alive when caught. After finishing its meal, the rat licked its front paws before reentering the water. Finally, before swimming away, it cut a mouthful of green grass for dessert.

Marshland lodge (Muskrat)

On nearly all drawings and paintings, the beaver is depicted gnawing a tree or working on its dam. In nature, such activities are rarely observed unless one makes a special effort and is prepared to sit and wait until it is unaware of one's presence. Wild beavers do not like to be watched while they are out of their watery element. I have painted the beaver the way one usually sees it — swimming, body low in the water, only its square head visible, leaving a long rippled wedge on the tree-shaded surface. The beaver holds a green poplar sprig in its teeth, perhaps its first bite of fresh food for five or six months since the ice closed it in and forced it to subsist on submerged stores of water-logged branches and bark.

One can well imagine the beaver's relief when it finally breaks free from its dark tomb, smells the spring breeze and feels the warmth of the sun. It is easy to become somewhat anthropomorphic where beavers are concerned. Indians spoke about the Beaver People, whose giant ancestors created the world by damming the flood. Like man, beavers manipulate their environment by physical toil and lay in stores of food. When I come across a deserted beaver lodge and examine the evidence of their labor, I am touched by nostalgic sentiment akin to the feeling one experiences when observing the ruins of a homestead or a decaying trapper's cabin in the wilderness.

Beavers influence the lives of many other creatures including birds, fish and mammals that find living space in or near the ponds created by the dams. Silt carried by the stilled stream sinks to the bottom and eventually fills the pond that may ultimately become a fertile garden for man.

A bite of greens (Beaver)

The weasels form an unusual and prominent family of carnivores that include such animal personalities as the otter, the marten and the badger, each of them a champion in its own specialized field. But the most outstanding character is no doubt the skunk. Quite unique among small mammals, it disregards protective coloring and openly flaunts its attributes, confident of its ability to repulse all but the most coarsely-tuned predators. However, use of its ultimate weapon is quite restrained and not easily provoked. I have approached dozens of skunks to within six meters, unwittingly or purposely, and none pulled the trigger. They just stood their ground, square-legged with raised tail, eyeing me cooly. Some even took a few quick steps towards me, sending me on a hasty retreat, mindful of the possibility that such aggressive behavior may be induced by rabies. Even when I teased a skunk and threw loose clods of earth at it, the brave animal did not push the panic button and exuded only a whiff of scent, which in minute dosage is not at all offensive.

Secure in their armament, skunks go about their business with the composure of a buffalo or grizzly bear, acting as if they did not have a care in the world. They are quite common along Beaverhill Lake, where they grub about in the pastures, turning over cow pads to look for insects, or amble along the muddy shore in search of carrion. I found one feeding on a red carpet of washed-up ladybugs; another scavenged from a dead porcupine, ending up with several quills in its nose.

One day in spring, I saw a magpie fly out over a marsh and land near another black-and-white object that turned out to be a skunk's tail waving over a tussock of grass. Walking up quietly, I stood and waited while the animal remained partly hidden in the tall grass. It appeared to be eating something as it was making the sounds of someone munching crackers. After the skunk ambled away, I examined the tussock and found a duck's nest with seven crushed eggs.

Skulking the shore (Striped skunk)

To the receptive wilderness hiker, a trip in the Rocky Mountains is an emotionally charged experience unattainable in other alpine regions of the Western world. The difference lies in the occurrence and reputation of the grizzly bear, whose powerful personality pervades the atmosphere of the Canadian mountains and brings modern man closer to his primitive ancestors. At nightfall, the lone hiker huddles by his fire, troubled by ancient memories of the beast that lurks in the dark, lusting for blood. In the bright light of morning, all worries are forgotten and the traveler feels reborn, filled with vitality. But when his trail leads through dense thickets along a roaring glacial stream, the muddy banks pitted with clawed tracks, he may look about him nervously and talk aloud to bolster his shaky courage. Where grizzlies are concerned, few people remain unaffected.

During my first visits to the Rockies my curiosity about bears bordered on foolhardiness. I followed a family of blacks into the woods, walking some distance behind to observe their behavior. The mother sent her cubs up a tree and stood her ground, ready to charge. A large boar allowed me to accompany him closely, until he went to sleep, or pretended to, behind an uprooted spruce. My confidence in the harmless disposition of all bears changed gradually as reports of mauled campers left their residue of fear. While hiking the back country trails, the risks of meeting a grizzly were uppermost on my mind. When I finally met the infamous beast, there was neither alarm nor terror involved. While her two cubs frolicked about, a big mama bear was peacefully grubbing for food on a flower-studded alpine meadow in Banff National Park.

I met other grizzlies in the Yukon mountains where the painting is set. The large boar was digging for ground squirrels, the width of his massive forearms awesome to see. He frequently rolled over for a nap, although he could not lie still for more than a minute. I had planned to camp in his meadow, but before dusk fell I hiked down the mountain to reach the road at dark.

Grizzly meadows

The fox approached along a shallow draw until it halted some ten meters away. A cold fire seemed to smolder in its opalescent eyes, their glittering glare upon me as I stood transfixed watching through binoculars. The next instant, the beautiful little barbarian flashed round and ran, streaming over the snow-covered stubble, its long tail floating after.

A close encounter such as this provides a glimpse into the fox's hunted life by conveying the intensity of its reactions when confronted with man, its adversary wherever the two exist together. In Alberta, the fox has another enemy, the coyote. Both species subsist on a similar range of food and are therefore competitors. The coyote has always been quite common on Alberta farmlands, especially on brushy pastures and other rough terrain away from human habitation, where it appears to have no use for a saucy little interloper such as the fox. Over the years, I have seen coyotes chase foxes on half a dozen occasions. The fox always proved to be faster and out-maneuvered its larger cousin, but to protect its pups the fox had to call on the involuntary assistance of its other enemy — man. To get away from the coyotes, the fox chose to live in suitable habitat around farms and along roads. Of course there was a certain amount of overlap; coyotes frequently invaded the fox's territory. With pups to worry about, both parents showed great anxiety and followed intruding coyotes closely, attempting to lead them away on a chase.

One evening, when I was sitting on the edge of a field some distance from a fox den, a coyote rushed up in a manner that did not promise much good for the pups which were playing or sleeping near the den. I held my breath, unwilling to interfere in nature's drama, but to my relief the vixen emerged from her hide-out and screamed a warning that sent the pups underground just in time. A few days later, I found the den deserted and picked up the tail of one of the young foxes from the mound. In the soft earth of the surrounding summerfallowed field, the only tracks were those of fox and coyote.

Living by its wits (Red fox)

Foxes are largely nocturnal in central Alberta. If they do travel during the day, they give evidence of an uncanny ability to remain out of sight, even in open country, choosing their runways so that the slope of the terrain hides them from hostile or curious eyes. Thus, they manage to exist near man without making their presence known to all but a few lucky or cognizant people, at the same time creating a challenge to the naturalist who is eager to learn about their habits.

In areas where they are seldom disturbed, foxes snooze the day away in the open or in a woodlot, summer and winter. Snowshoeing along their trail of the night before, I may come across their bed — a circular depression about 30 centimeters across, a few reddish hairs clinging to the iced surface, but the animal is gone. Pushing on along its fresh tracks, I try to guess where the fox will lie up next and may reach the spot after the fox has bolted again, all the while staying out of sight. In localities where foxes occur, I have learned to look for them by snowshoeing or skiing against the wind, scanning the terrain ahead through binoculars, scrutinizing the lee side of every slope or knoll. The fox will lie up near the top, protected from chilling breezes, with a good view to one side, while its nose and ears warn of dangers coming from the upwind direction. Upon discovery of a sleeping fox, I advance slowly, ready to freeze as soon as the animal raises its head.

As long as I remained motionless, some foxes did not seem to recognize my hulking human shape. On a wintry afternoon when the molded drifts along a ravine were reflecting the orange glow of the setting sun, I located a large male fox, sleeping so tightly that I could approach to 50 meters. He got up once, gave me a drowsy glance, stretched his black velvety legs and pirouetted in his bed, going round twice before finding a position that satisfied his idea of comfort. Then he curled up snugly again, his brush settling over his nose like a baby's quilt.

On the drifts (Red fox)

During summer foxes eat a variety of foods. Besides preying on mammals and birds ranging in size from mice and sparrows to hares and mallards, they nibble quantities of insects, frogs, berries and other vegetable matter. For the northern plains fox the menu list shrinks considerably in winter, when very few animals manage to cope with the harsh conditions on the frozen fields.

Particularly after heavy snowfall, I follow the fox's trails with interest as well as concern, but there is no need to worry; "Reynard" is a superlative hunter that thrives as long as there are mice and voles. These little rodents find abundant food in hay meadows and stubble fields, while snow acts as an insulating blanket under which they lead their high-strung lives, unaware of the fox that stalks gingerly, its nose close to the ground. Whether its nose actually plays the dominant role in locating this type of prey is hard to know; it is certain that the fox's black ears are ever active, decoding audible signals from far and near, accurately gauging the distance to their source. Suddenly the fox stands to attention, head cocked in a pose of intense concentration. There he goes! In a high flying leap he springs three to five meters away, landing on all fours, pinning the vole down and seizing the prey with its teeth in an instant. The fox can put up a show that is both comical and dexterous, pouncing repeatedly and prancing about, tail swinging as a rudder. In height and horizontal distance of its lunges the fox easily out-performs the coyote, which also hunts mice and voles in this way.

In late winter when the drifts deepen and the crust hardens so that a man on snowshoes stays on top, the fox's foraging becomes more difficult. Its pounces are nearly vertical and it hits the snow front feet first, maximizing the impact in an attempt to collapse the tunnels through which mice and voles travel. Digging furiously, the fox tries to uncover the rodent, thrusting its jaws deep into the snow. Coming up with the prize in its mouth, the fox's arched pose is like an athlete's raising of the trophy!

Following its nose (Red fox)

Reynard is a compulsive hunter who seldom passes up a chance to catch a prey that happens to present itself in a vulnerable position. He may even interrupt his sleep to snatch up a passing vole or mouse. If it is not consumed on the spot, the prey is stored until needed and buried into soil or snow. With its front feet deftly alternating, the fox quickly digs a hole large enough to hold the food item, covering it up by pushing the loosened material into place with a few strokes of the nose. Such a cache keeps the meat cool, away from egg-laying flies and out of sight of scavenging crows, jays, magpies, ravens or hawks.

When a traveling fox has caught a vole or mouse that it wants to cache, it moves off to look for a suitable spot often marked by a tussock or other slight prominence serving as a marker. When the fox comes this way again, a routine inspection of all points of interest may lead to the uncovering of such a cache at a time when hunting is poor and the fox is hungry. During a fox's travels, its sensitive nose may locate several filled caches, but also numerous empty ones as well as inedible food remains such as bits and pieces of bone and skin buried under snow. The fox might waste a lot of its time investigating such unproductive spots, were it not for its "bookkeeping system". Empty caches and old prey remains are marked with a few drops of urine that signal "No food here! Go on" to the next fox that may come along. This unique bit of new knowledge about the red fox was discovered by David Henry who studied the species in Prince Albert National Park, where some foxes had lost their fear of man and allowed close observation. On the farmlands, I had to unravel their habits mostly by following tracks in snow. Upon coming across signs of hunting activity, it was a challenge to locate the cache, if there was one. A disturbance in the snow along the trail, ever so slight, made me run a gloved hand several centimeters under the surface. It was a thrill to uncover a prey item, usually a short-tailed meadow vole, a long-tailed deer mouse, or a pointy-nosed shrew.

The cache (Red fox)

Epilogue

To the reader, going through these pages may have been like spending a day in nature; one could not be sure which bird or mammal would show up next. In the field, wild animals are seldom met with in a predictable pattern, never in the systematic order established by the academic. The field-naturalist rarely observes a complete group of creatures that share a certain habitat. Such completeness and the creation of an "unnatural" order are of value only in the contents of handbooks and guides. A book which features the author's personal experiences with birds and mammals may be compared to a collection of short stories; their length and subject are purely a matter of the writer's choice and inspiration. For reasons given in the introduction, these paintings are dominated by certain favorite animals. If you missed some of your favorite species, there may be another chance in a future volume.